591.5 Pettit, Ted S.
PET
      Wildlife at night

DATE			
10-12			

© THE BAKER & TAYLOR CO.

ABOUT THE BOOK

Wherever you live—in city, suburb, or open country—
there is fascinating wildlife to observe at night. In this
book the naturalist and writer Ted S. Pettit shows how
watching this night life can be more fun than what you
may see during the day. With interesting and amusing
true stories of his own experiences and those of others
he reveals the wonderful night world populated by animals, birds, fish, and insects and offers practical suggestions on how you can see it in action.

Wildlife at Night

by Ted S. Pettit

G. P. Putnam's Sons New York

To Susie

Whose interest in wild animals is confined largely to her father's income derived from writing about them.

Copyright © 1976 by Ted S. Pettit
All rights reserved. Published
simultaneously in Canada by
Longman Canada Limited, Toronto.
SBN: GB-399-61026-X
SBN: TR-399-20536-5

Library of Congress Cataloging in Publication Data
Pettit, Theodore S
Wildlife at Night
Includes index.
1. Nocturnal animals—Juvenile literature.
2. Wildlife watching—Juvenile literature.
[1. Nocturnal animals. 2. Wildlife watching.
3. Animals—Habits and behavior] I. Title.
QL755.5.P47 1976   591.5   76-13596

Contents

1 Night Life of Wildlife

It was well after midnight, and my mother, father, brother, sister, and I had been asleep for several hours. Suddenly a most unusual noise awakened us. The noise came from under the kitchen floor and sounded as if someone were using a miniature jackhammer to pound a hole in the floor.

That was our first night in an old house in the mountains that my father had bought and fixed up as a summer and weekend vacation retreat. The closest neighbors were a half mile away. There was no electricity in the house that night, and as the family gathered in the kitchen to investigate the weird sound, each of us had a flashlight. The noise stopped.

My father suggested we be quiet and turn off the lights. In a minute or two the noise started up again, louder than before, and we could feel vibrations coming through the floor as we stood there in our bare feet. My brother, who was five, started to cry, and my sister dived back into her bed and hid under the covers. My mother stood there motionless, not knowing what to do.

"Put on your shoes," my father told me. "Let's see what it is." We put on shoes as quickly as we could and, with flashlights in hand, rushed out the front door and around the house to the cellar entrance. Silently we went down the stairs, opened the door, which was ajar, and aimed the lights at the cellar ceiling where we thought the noise had come from. There, on a wooden post which supported the ceiling beams, was a full-grown porcupine. It had been chewing on the underside of the kitchen floor. When a porcupine chews, we discovered, it chatters its teeth.

That was my first experience with wildlife at night, but not the last.

Some years later my family and I spent a spring weekend in the same house. We had arrived on a Friday night, parked the car in front of the house, and did not pay much attention to it until Sunday afternoon. Then we loaded it up for our return home, and my daughters climbed into the back seat. At that point a loud bang came from the back of the car that could mean only one thing—a blowout. It did not take much detective work to see what had happened.

A porcupine had been chewing on the side wall of the tire, severely weakening it. The weight of my two daughters did it. The tire blew. We checked the other three tires and found tooth marks on one other, but not deep enough to be serious.

After we had put on the spare and driven to the service station to get a new tire, the mechanic immediately put the car on the rack and began inspecting the brake lines. Porcupines, he told us, sometimes chew through the brake line to get at the brake fluid. Only the week before a man had run into a tree when his brakes failed owing to porcupine damage.

Porcupines are active mostly at night. During the day they climb into a tree to sleep or crawl into a cave, under a rock, or a fallen tree. Naturally they eat tree bark or twigs and buds. But they have a great craving for salt and, in trying to get it, do a great deal of damage. The chemicals used in tires and brake fluid contain salt.

Many a camper or woodsman has found this out the hard way. Axes or canoe paddles, left out where porcupines can get at them, have been ruined in one night. Perspiration on ax handles or paddles contains salt and is very attractive to porcupines.

The superintendent of grounds at a surburban New Jersey office building had his experience with wildlife at night in a more amusing way.

When the office was built, a reflecting pool had been constructed at the main entrance. Pond lilies and other plants were placed in the pool, along with a half dozen medium-sized goldfish. All went well for a week or two. Then the goldfish disappeared. The groundskeeper was puzzled by the disappearance but bought some new fish, which he placed in the pond. When these too were gone, he became more annoyed than puzzled and decided to see what was going on.

He bought some new fish and that night hid near the pool until about ten o'clock. When nothing unusual occurred, he went home. The next day those fish were gone. Now thoroughly angry and bewildered, he replaced the fish, and rigged up a dim light that shone on the pool. He placed a chair in some dense ornamental shrubs near the pool where he would be hidden but could still see the pool. He was determined to stay all night if necessary.

Near midnight he was aroused by a strange noise overhead. It sounded like a loud *quack, quack, quack*. Out of the darkness flew two black-crowned night herons, long-legged and long-billed birds about fifteen inches high.

They landed on the edge of the reflecting pool and looked into the shallow water, and then each neatly grabbed a goldfish and swallowed it. In about ten minutes all the fish had been gobbled up as the groundskeeper watched, too amazed to frighten the birds away.

Night herons usually feed in swamps, marshes, lake shores, or shallow streams. They eat small fish or frogs. But these New Jersey birds knew a good thing when they found it. Why fly all the way to a river when a kindly groundskeeper kept supplying them with food? The groundskeeper abandoned the idea of fish and concentrated on pond lilies.

A woman in a town in Westchester County, New York, once became involved with wildlife at night, much to her consternation. The women's hats of the day were small, round fur pieces which were worn on the top of the head. One night, as the lady was leaving her church to walk to her car, something grabbed the hat off her head and disappeared. The next day her hat was found on the church lawn.

At first mischievous boys were blamed for the incident. They had climbed up the tree over the sidewalk, it was said, and reached down to snatch the hat.

But the real culprit, it turned out, was a barn owl that nested in the church steeple. Barn owls feed at night, largely on rats and mice. As the name implies, they sometimes nest in barns, but they commonly nest in church steeples, attics of abandoned houses, and similar places.

The point of these stories is that wherever you live—in city, suburb, or open country—there is fascinating wildlife to observe at night. In many cases it is much more interesting and fun than what you may see during the daytime.

For there are many more animals active at night than in the daytime, and some are active both day and night. Why animals are active and what they do are fascinating. You may never suspect that animals are there, but a little detective work will show an unbelievable number doing many interesting things. Here you can learn what some of these animals are and how you may find out what they do.

But first what do we mean by animals, and what do we mean by night?

The animal kingdom is divided into two main parts: animals without backbones, called invertebrates, and animals with backbones, called vertebrates.

The invertebrates are insects, spiders, crabs, crayfish, mollusks, and worms. Vertebrates are snakes, toads, frogs, fish, birds, and mammals. Within each of these groups are animals that are active only at night or both day and night.

"Night" is the time between sunset and sunrise. Night-time may be longer at some times of the year than at others or longer or shorter in some places than others. The length of time between sunset and darkness also differs according to time of year or place or a combination of the two.

2 Mammals at Night

The nighttime activity of a mammal had caused trouble for our neighbors and us for several weeks. At first we thought it was a dog that overturned garbage cans each night and spread the trash all over the yard. But then we found the tracks of a huge raccoon in the soft earth of our garden and knew who the real culprit was.

We had tried to capture the animal in a box trap, baited with sardines, and failed. The raccoon was too large for the trap. We had used strong springs to fasten the top securely on the can, but the raccoon was strong enough to open the can anyway.

Other mammals are active in our yard, too, mostly after dark. We rarely see cottontail rabbits before dusk, but their tracks and chewed leaves on beans and lettuce in the garden show they are there. Fortunately a low fence and mothballs spread around now keep them out of the garden.

One summer we had a real mystery. In the middle of the lawn was a birdbath we cleaned and filled each evening so that it was ready for birds the first thing in the morning. But one morning the birdbath was dry. We checked for a leak but could not find one. We filled the bath, and it held water all day.

The next morning it was almost dry again. This went on for several nights, and we tried to figure out what was going on. Then, one night about eleven o'clock, we heard a noise in the back of the house and flipped on the outside floodlights. There at the birdbath were two deer. The mystery was solved. It was a summer with little rain, and nearby streams were dry. The deer had been coming to the birdbath for a drink.

We had a somewhat similar experience when we lived in a third-floor apartment in Forest Hills, New York. There were blue jays, sparrows, and other birds in the area, so we put a bird feeder on a windowsill. Each evening we filled it with sunflower seeds and a few peanuts. Birds came during the day, but then we noticed something strange. Something was also coming at night to take away the peanuts. We knew that birds and squirrels are not active at night, but what was taking the peanuts? I decided to set a camera trap with a flash bulb to find out. Within ten minutes that night after we rigged up the camera, the bulb flashed. Something had moved a peanut that worked the trigger.

I quickly changed films, replaced the bulb, and reset the trigger. In about five minutes the second bulb flashed. At that point I removed the film and rushed to my darkroom to develop it. In a few minutes I had the answer. White-footed mice, which apparently lived in the ivy on the front of the house, had found the feeder and were stealing the peanuts.

Like the raccoon, deer, and mice, many other mammals are active at night. Some, such as bats, are active only at night. Even in the city bats may be seen darting under streetlights to catch insects on which they feed. Bats are the only mammals that truly fly. Flying squirrels, also active at night, can only glide from a high point to a lower point.

Bats

Bats are unique among mammals in other ways. As they fly, they emit very high-pitched calls, too high for the human ear to hear. These calls bounce off insects or obstacles, and the echo is picked up by the bats' ears. Thus, bats locate insects in the air. But the echoes also warn the bat of obstacles in the flight path, so they can be avoided.

If you are lucky enough to see bats at dusk and they are swooping not too high in the air, or if you see them under a streetlight, try this stunt for a closer view. First, collect a handful of small pebbles about the size of peas, or use dried peas. As a bat approaches you or the light, toss one of the pebbles up in the air. Many times the bat's "radar" will pick up the pebble, and the bat will follow the pebble right down to the ground. Some naturalists have used this trick to capture bats in a net for closer study.

Bats can squeeze through remarkably small openings and sometimes get into houses. This may cause considerable excitement because many people still think that that bats will get in their hair. That is about the last thing a bat wants to do. All it wants as it flies around the house is to get out where it can find insects.

Little brown bats have found ways of getting into

three houses in which I have lived, one house being right in the middle of town. One time we had three bats at once flying around the bedroom. The wind from their beating wings woke us up. As we turned on the light, the bats did the best disappearing act we ever saw.

But they soon appeared again to fly around the room. Fortunately we had a lightweight nylon insect net handy and soon learned how to net the bat as it flew by. The bat could avoid the net if it was in front of it, but a quick backhand swipe as it flew past usually worked.

Bats have very sharp little teeth, so it is best not to handle them unless you know how.

The eyes of some mammals that are active chiefly after dark have a kind of mirror in the back that reflects light. The light intensity at night is so low that this layer is necessary for the animal to see at all. When a bright light, such as a car headlight or flashlight, is shone into these animals' eyes, the eyes glow much like the reflection on the back of bicycles.

A night hike along a country road with a bright flashlight frequently results in your seeing animals that you might otherwise miss. The color, size, and location of the eyes give you the clue to which animal you see. Walk along quietly and slowly, shining the light along the ground some distance ahead and all around. Aim it at trees, at stone walls, and along stream or lakeshores. If you are on a lake and have a boat, row quietly about one hundred feet from the shore and parallel to it. Aim the beam of the light along the shore. Many times it is possible to pick up the eye shine of deer or raccoons as they come to drink.

(Animals other than mammals, such as some birds, spiders, frogs, and crocodiles also glow. They will be discussed in other chapters.)

Eye Chart

Bright orange eyes, close together	Bear
Bright yellow eyes	Raccoon
Bright white	Dog, fox
Yellowish white	Bobcat
Bright white, up in a tree	Porcupine

Bears

Anyone who has visited or camped in national parks or forests and some state campsites has probably had experiences with bears. They can be scary experiences or not, and many are very annoying. But with certain precautions you can learn to live with bears, as millions of campers have, and take the bad with the good.

Our first experience with bears was the first time we camped in Yellowstone Park. A naturalist friend had told us we would be happier staying in a park cabin rather than in a tent. We arrived in the park early in the morning and were lucky to get a cabin.

As we walked into the small one-room cabin, we wondered why the windows were small and high on the wall so that we had to stand on the bed to look out. We wondered too about the heavy, reinforced door without windows and with a very heavy lock. That night we found out.

We spent the day fishing and sightseeing and got back to the cabin just before dusk to cook dinner. Halfway through dinner we heard a racket in the cabin and jumped up on the bed to look out the window. There in the light of a streetlight we saw was a huge bear hauling cans, bottles, and other refuse out of the garbage pit while searching for food.

This was not an ordinary garbage can. It was a hole in the ground lined with concrete. The top was heavy iron that could be opened only by stepping on a treadle a few feet from the top. It took two of us to open the top—one to step on the treadle and the other to dump garbage in the pit.

The bear, however, had learned to open the top alone, then to reach in for food. There was a real mess around the pit by the time the bear looked for other places to feed.

Bears, like most animals in the wild, eat what is easiest to get, and in many parks until recently garbage was the easiest to get.

Later that night my daughters, wife, and I walked to the washrooms which were about one hundred feet from the cabins. The men's washroom was on one end and the women's in the other end of the same building. My wife and daughters made me promise I would meet them there and we would walk back to the cabin together.

As they were waiting for me, another camper came along and asked what they were waiting for. When she was told, she answered, "Oh, I thought you were waiting for the bear. Last night, when we came here, there was a bear in the ladies' room."

A few years later we went back to Yellowstone and again stayed in a cabin. The first night we went to the evening campfire program in an outdoor auditorium, where naturalists give interesting talks on the natural history of the park.

That night the naturalists opened his talk by saying there was a bear in the park that was very much interested in natural history. If the auditorium was only half full, the bear sat in the last row and watched the slides. Of course, every head turned to look, and sure enough, there was a bear sitting on its haunches watching the show.

Unfortunately, too many national park campers think the bears are tame or have been trained to entertain visitors. The truth is that the bears are wild, although some have lost their fear of man. Too many people have disregarded park regulations and lived to regret it. Some have been mauled by bears; some have had the doors torn off their trailers or campers; some have had their tents destroyed or their cars damaged.

When park regulations prohibit feeding bears, it is for good reason. When they recommend not leaving food in the trailer, it is for good reason. On one visit we carefully removed all food from the car and took it to the cabin. Even then one night a bear severely scratched the trunk of the car, attracted by the smell of the food that had been there. If all regulations were observed, as well as safety suggestions, there would be few man-bear confrontations.

Man-bear relations are not restricted to national parks. We once arrived at a well-populated campsite just outside Ottawa, Canada. We immediately noticed a pile of new cans of fruit with two neat little holes just below the rim. The holes looked as if someone had shot at the cans with a .22 caliber rifle. But both the punctures went into the can, not one in and one out as would happen with a bullet. Bears had learned to bite a can and drain out the sweet syrup, which they lapped up. The night before our arrival they had almost cleaned out one camper's supply of food.

A hunting club owns an old schoolhouse not far from our farm, and some members learned the hard way how destructive bears can be. A group hunted all one weekend and went home on Sunday, to return the next Friday. They left a side of bacon hanging in the cabin along with an assortment of other food.

One night in their absence a bear smelled the bacon and tried to get it. In the process it tore a hole in the side of the building instead of going through a window, ate the bacon, and went out a window on the other side, enlarging the window opening in the process. We were long ago advised not to leave such food in our farmhouse, and we do not.

Bears are not necessarily animals of the wilder parts of the country. They have been seen in some of our larger cities, Minneapolis, for example, and even rather close to New York. In the summer of 1975 a bear or bears were reported seen about forty miles from Times Square. No one knows for sure whether it was the same bear in Somerset County, New Jersey, but there were several sightings, one at an early-evening garden party.

One night, while driving through the Delaware Water Gap, I had to come to a screaming halt to avoid a bear stretched out in the middle of the road. I thought it might be injured, but I wasn't about to get out and see. With my headlights on the bear, I blew the horn. Nothing happened. After about five minutes of horn blowing and inching closer and closer, the animal stood, stared straight into the headlights, and ambled away.

One fall a few years ago a bear visited a farm not far from our home each night and gradually wrecked a garden. It also climbed fruit trees to dine on apples and pears, much to the chagrin of the landowner. You can hardly blame him for shooting the animal when he found it asleep in an apple tree. But it was contrary to law, and he paid a fine.

Most bears, though, are not destructive animals since they feed on roots and shoots, berries, nuts, and small animals. But they do eat what is easy to get, and if they find an apiary, they will wreck the hives to get the honey. If they find an untended flock of sheep, they may develop a liking for lamb.

However, these are isolated cases, and bears deserve continued protection as one of our larger and interesting wild animals.

Beavers

One of our more enjoyable experiences with wildlife at night, or at least at dusk, took place one summer watching beavers. The year before, beavers had damned the stream on our New York State farm and had built a huge lodge on one bank. We discovered we could hide behind a large elm tree on the bank and watch the beavers as they came out at sunset to work on the dam or feed on aspen trees that grew nearby. There was a period of about half an hour between sunset and darkness during which we saw beaver activity very well.

The pool formed by the dam was about one hundred

feet long, twenty feet wide, and six to eight feet in depth at the deepest point. On the bottom of the pond were branches of aspen which the beavers had poked under rocks so they would not float away.

The first thing we noticed was that the entrance to the lodge was underwater. When the beavers came out, they swam underwater a good distance from the lodge before coming to the surface. Then only their heads were above-water as they looked for any possible enemies. Seeing none, they would dive for aspen sticks, carry them to the stream bank, and start to feed on the bark.

In feeding, they held the stick with their front feet and rotated it, looking for all the world like a person eating corn on the cob. After feeding in this way for a few min-utes, the beaver would shred bark from the branch with-out eating it. When it had a good-sized mouthful, it would swim to the lodge, dive into the entrance, and take the bark to its young. We could hear the young inside the lodge, mewing like kittens.

Meantime, the other adult beaver swam upstream and climbed out on the stream bank of the pool. We did not see the young until about midsummer. Then each evening three young came out out of the lodge to swim around and exercise. The adults would give them aspen sticks, and occasionally we could watch the whole family feed-ing on the stream bank.

When it was so dark we could barely see, we would try to back away quietly in order not to disturb the beav-ers. But occasionally we would kick a loose stone that would slide down the bank into the water. Then we would hear the beaver's alarm system. One or both of the adults would slap the water with their broad tails and a loud noise would send the young swimming back to the lodge.

Once we wanted to see how the beavers worked on the dam. Late in the afternoon we removed a few sticks and some stones and mud so the water rushed through. We had been told the sound of the water rushing through the hole and the drop in the water level in the pool would bring out the beavers to make repairs.

At dusk, in our usual hiding place, we found that our stunt was successful. One beaver was at work swimming back and forth, taking new sticks about the size of a broom handle to plug the hole. Occasionally it would dive to the bottom, get some mud in its front feet, and pack the mud in between the sticks. In a remarkably short time, the repairs were completed.

The amount of work beavers can do in one night is amazing. One summer we were camping in northern Minnesota on a site bordered by a small brook. Beavers moved in one night and started a dam. The first night the dam was about six inches high and six feet long and effectively backed up a small pool. Nobody thought much about it until the third night when the dam was about two feet high and the stream started to overflow its banks and flood the road.

The camp ranger fastened a large grappling hook to a chain behind his tractor and pulled out about half the dam. That night the beavers repaired the damage. This went on for several days and nights until finally state conservation department people live-trapped the beavers and moved them to another stream miles away.

Another summer we were on a fishing trip in Ontario

when some other fishermen who had never seen or heard a beaver learned about them the hard way. The first evening at sunset they left their camp and chugged down the lake to a small cove to fish for bass. But they quickly returned and complained to the conservation officer who was in charge of law enforcement on the lake. Someone, they said, was hiding in the bushes on the lakeshore and throwing flat rocks out in the lake. One rock came so close it splashed water on them.

It took awhile to convince them that they were anchored near a beaver lodge and that the beavers had learned that if they smacked the water with their flat tails, boats would go away. Fishermen learned that with all the noise created by the beavers, fish would not bite anyway.

William Carr of Tucson, Arizona, told us an amusing beaver story. He had an idea for a beaver museum to be constructed in New Mexico that would include a man-made indoor-outdoor beaver lodge so that visitors could see beavers both in their lodge and feeding in a pool. The lodge was made of concrete exactly the same as beavers build them of sticks and mud. The inside of the lodge had a shelf above the water level where the beavers could sleep, and there was an underwater tunnel to an outside pool. The lodge itself was inside a building that housed a beaver library and some exhibits.

There was a window through which visitors could look into the lodge and a door that caretakers used to enter the lodge to clean it. One evening the caretaker did not close the door securely, and that night the beavers pushed the door open and got into the library. They were possibly the first beavers in history to dine on wood paneling and table legs.

Wildlife at night can create some scary situations, as happened one summer in New Hampshire at a Boy Scout camp I was visiting. It was about 10:30 P.M., and several of use were sitting around a small campfire swapping stories. Suddenly all of us jumped up as a weird scream came from the woods. It could have been a boy in agony. Three of us ran in the direction of the noise, while the camp director ordered a bed check to see who was not accounted for. We searched the woods for half an hour and found nothing. Then the camp director reported that all the boys were accounted for.

The next day we found the source of the bewildering scream. A raccoon had fallen into a half-filled cistern and was unable to climb out. Apparently it swam until exhausted and screamed just before it drowned.

Skunks

Although seen during the day, skunks are generally more active at night, a fact that two hikers I once met will never forget. With two friends I was climbing a mountain in the Catskills in New York State when we came to a log lean-to built to shelter hikers. Knowing there was a cold spring nearby, we left the trail to fill our canteens. As we approached the lean-to, we heard someone say, "Sh, Sh, Sh."

We stopped talking when we saw two hammocks slung across the inside of the lean-to, each occupied. Since it was about noon, that seemed a little strange. Then a voice said, "Is it still there?" I asked, "Is what still there?"

"The skunk that is eating our food."

We backed up, naturally, not wanting to be the target of the strong scent that is the skunk's defense weapon. It was obvious something had been eating the food, but we did not see a skunk. Then we heard the story.

The hikers had arrived at dusk the night before, slung their hammocks, and cooked dinner. They then put the balance of the food in a pack and put the pack on the ground under the hammocks. They had just about climbed into bed for the night when they heard something rumaging in the food supply. One of the hikers pointed his flashlight at the pack, right in the face of a skunk.

For obvious reasons, he pulled back, whispered to his companion, and both remained quiet, hoping the skunk would go away. It did, but they didn't know it. In dragging the food out of the pack, the skunk loosened some paper that blew in the wind. Hearing the paper, they thought the skunk was still there. Not wanting to be on the receiving end of a skunk's famous weapon of defense, they stayed in bed till noon.

I once had a strange experience with a skunk, or it could have been skunks. I'm not positive it was the same animal I saw twice, but suspect it was.

The first evening I fly-fished up my favorite trout stream until almost dark. I climbed up the bank to the dirt road and started walking back home, about a mile away. Having a strange feeling that someone was watching me, I turned around. There in the road was a skunk looking at

me, but not on the defensive. I walked a little faster, looking over my shoulder, and the skunk walked a little faster.
I stopped, and the skunk stopped. I started up again, and
so did the skunk. I don't know how far it followed me
since by then it was dark. Exactly the same thing happened three nights later in the same place.

A neighbor of ours had an interesting experience with
skunks in the middle of a suburban New Jersey town.

One day she had been cleaning her basement when she
forgot to close the outside door. That night a skunk got
into the cellar, a fact she discovered unhappily the next
morning. She called the conservation officer to ask what
to do.

"That's easy, lady," he told her. "Leave the door open.
Break up some cookies, and make a trail from the basement up the stairs to the outside. Tonight the skunk will
follow the trail as it eats and will find its way out."

But it didn't work that way. The next morning there
were two skunks in the basement. But that night they
finally did find their way out.

In another part of that same town a family of skunks
lived in harmony with man for a year or more. We could
never be sure where they found shelter, but we saw them
several times during the year, feeding at night from
garbage cans near the county administration building.

Opossums

The opossum is another common mammal in some
parts of the country that is seldom seen except at night.
It has been seen in the middle of New York City and

other cities, and many a suburban dweller has had to clean up after a possum pushed over his trash can to get its nightly meal.

Most people see possums in the headlight beams of the car as the animals scurry across the road at night. Many more are seen dead in the road when they were not fast enough to beat the cars.

The opossum is now found over a larger area than it was only a few years ago. This animal is the only one native to the United States that carries its young in a pouch. It is found from New England west to Minnesota, south to western Texas, and throughout the Southeast. It is also found in parts of southern California.

Possums are about the size of a house cat but have a chunkier body, shorter legs, and a long pointed nose. They are usually gray but may be any variation from white to black. Their tails are naked, and their ears are black. Opossums are usually active at night.

Possums eat almost anything from fruit and vegetables to chickens or dead animal matter. They are trapped for their fur and are hunted in some places for sport or food. They live wherever they can find shelter, from hollow trees to burrows or holes under barns or farm outbuildings. Up to fourteen to sixteen young are born at one time, but usually only seven or eight survive.

When born, opossums are about one-half inch long, with a tail an inch or more in length. They live in their mother's pouch for seven or eight weeks. Young opossums are seen frequently clinging to the back of the female. The possum is famous in folk songs of the South.

If there are possums in your area, you may be able to

see them by using the "Lighted Tracking Pit" trick described later.

Deer

The best time to watch for deer is at dusk and after dark with a bright spotlight. Since many times deer have regular feeding and travel habits, a little scouting around during the day helps to find deer at night.

Look for deer tracks in a well-worn trail in woods or across fields. Look for shrubs or small trees with the tips of the twigs chewed. You may find one place where deer drink from a spring or stream. In the fall look for old apple trees with partly eaten apples on the ground.

When you find clues that show deer might be present, find a spot a couple of hundred feet away where you can hide. But check the wind. Hide in a place where the wind blows from the deer toward you. Once you get the deer in the beam of the light, hold the light steady. You should be able to watch the deer for a minute or two.

The deer seems to be dazzled by the light and will stand and stare at it for some time. Some hunters use this method to shoot deer in spite of the fact this is against the law. If caught, they could wind up in jail or, at the least, get off with a heavy fine.

If you are lucky enough to come across a deer at dusk and the deer sees you but isn't immediately frightened, try this trick. Step behind a tree, or if there is no tree nearby, stand perfectly still. Slowly, with no sudden motions, carefully get your white handkerchief and hold it by one corner. Flick it once so that it resembles the deer's tail as it twitches while the deer feeds.

If all goes well, and the wind is in the right direction, the deer will walk toward you. Using that stunt, I have lured deer to within six to eight feet of the tree I was standing behind.

Another clue to look for that will show deer are in the area is a spot where they have bedded down for the night after finishing feeding. Look for a place where the grass is matted down in an area two or three feet square. It may be in an open field or an opening in the woods. If you find it early in the morning, it may be still warm from the animal's body heat.

We once decided to have some fun with people who drove through our farm looking for deer with powerful spotlights on their cars. We painted a board eight inches wide and three feet high dead black. Close to the end of the board we mounted two one-inch yellow bicycle reflectors about six inches apart. We carefully stood the board on end, propping it against a tree about three hundred feet from the road. But behind that board we had another with one red and one green reflector in the same position as the yellow one on the front board. A long piece of string ran from the front board to our back window.

When the spotlighters hit the yellow reflectors, they stopped the car to watch. We gave them a few seconds, then pulled the string. Suddenly the "deer's" eyes changed color, and the exclamations of bewilderment that came from the car were most interesting.

Flying Squirrels

Rarely seen, but surprisingly common within their

range, flying squirrels are interesting animals. They are about ten inches long, including four-to five-inch tails. Stretching from their front to hind legs and along their sides are folds of fur-covered skin. They are brown on their backs and white underneath. When stretched out, they serve as "planes," enabling the animal to glide from a high point to a lower place. They cannot fly, though, as does a bat or bird.

Flying squirrels are found over the eastern half of the country, in parts of the Northwest, California, Idaho, Utah, and north into Canada. They live only in forested areas and have their dens in hollow trees. They sleep during the day and come out at dusk to feed on nuts, fruits, some insects, and other small animal materials.

Flying squirrels may sometimes be seen on bird feeders at night if a bright light is suddenly turned on. They make interesting, delightful pets and tame easily.

We do not know how long a flying squirrel lived in the stub of a half-dead maple in our backyard. We have only seen the animal twice. The first time was at dusk. I glanced up in the tree at a robin singing and saw the squirrel crawling around the branches, eating buds that were just about ready to break into leaf. The second time was a few days later.

I tried pounding on the bottom of the tree trunk with a stick to see if the squirrel would poke its head out, but nothing happened.

Mice

Most mice are lively only at night, but some are active both day and night. They are our most abundant mammals, and there are very few places in the country where at least one species does not live.

Deer mice, or white-footed mice, are found over most of North America. They are long-tailed mice, some being as long as seven and a half inches, including the tails. They have large ears, large eyes, and all are gray to brown on the back, white underneath, with white feet. These mice are active at night, coming out of the nests at dusk to feed.

Deer mice sometimes take over birdhouses or old bird nests or squirrel nests for their own. They also build their own nests, in hollow trees, in stumps, under logs, or even in houses or barns. They frequently move into summer cottages and camps and may be destructive of mattresses, pillows, clothing, and food.

Captured deer mice make delightful pets and are among the most interesting of the small mammals to watch. As winter approaches, they store up large supplies of seeds, nuts, and other foods for use during cold weather.

Deer mice may have two or three litters of three to seven young a year. The young grow quickly and in scarcely more than a month may be ready to have young of their own.

Jumping mice are small mice, seven to ten inches long, including long tails. They have long hind feet. Their tails are almost twice as long as their bodies. There are two kinds—meadow jumping mice, which are yellowish brown in color, and woodland jumping mice, which are yellowish on the sides, brown on the back, and white on the tip of the tail.

BAT

FLYING
SQUIRREL

BEAVER

PORCUPINE

SKUNK

RACCOON

Both mice live in burrows in the ground where they build a grassy nest. They spend the winter in hibernation in this burrow.

Jumping mice are found over most of the country north of a line from Georgia to southern California. They are active at night and feed largely on plant materials. Occasionally they may be found during the day under logs or rocks or along hedgerows.

When disturbed, they flee with long jumps, some as long as six or eight feet. They may be attracted with food as bait, but do not live well in captivity.

Bog lemmings are small mice which are about five inches long, including short tails of less than an inch in length. They are found across Canada and in the northeastern United States from Minnesota to Missouri, east to Virginia, and north to Maine. Their fur is rather long and is brownish in color on the back and lighter underneath. The tail is dark on top, light underneath.

These mice live in small colonies in damp woods and meadows. In wooded areas they make runways or shallow tunnels in the leaf mold on the forest floor. In meadows and fields they clip down the grass to make runways. They feed on green plant materials such as grasses and on some seeds. They build nests in the ground, made of grasses and leaves, and have several litters of young each year. They provide food for foxes, snakes, hawks, owls, and weasels.

Collared lemmings are found only in the Arctic. They have heavy, long fur and short tails. Collared lemmings are the only rodents that turn white in winter. They are buffy gray in color in the summer.

These animals are famous for their mass migrations. Hundreds of thousands to millions of them sometimes move for long distances to new areas. In the process of these moves they may come to water, and so great is their instinct to migrate that they all drown.

They are important in the Arctic as sources of food for furbearing animals and other predators. When the numbers of lemmings are low, the furbearers starve, and the trappers, too, have a difficult time. It is believed that when lemmings are low in numbers, snowy owls are forced south from the Arctic and Canada into the northern United States to find food. These animals are active day and night, winter and summer, and, unlike other Arctic animals, have young all year round.

Voles of one kind or another are found over most of the United States. The most widely distributed of these animals are the meadow mouse, the red-backed mouse, and the pine mouse.

Meadow mice are found in the northern half of the country from Idaho to New Mexico, east to the Carolinas, and north to Maine. As the name implies, they prefer fields and meadows, where they make runaways in the grass and build grassy nests in protected places. They may well be the most abundant of all small mammals, since in some places tests have revealed more than two hundred per acre, and in other places several thousand. They eat large quantities of plant materials and because of their numbers are tremendously destructive of food crops. They may have twelve litters of four to six young per year, and these young may have young of their own in a little over a month.

Meadow mice are five or six inches long, including one-to two-inch tails. They are grayish in color, and the tail is dark on top, light below.

Pine mice are found east of a line from Texas to Wisconsin, except in southern Florida. They are four to five inches long, with short tails. They have rich reddish brown fur and short ears. Despite the name, they live in hardwood forests, where they burrow in the litter on the ground and in soft earth underneath. They build nests in shallow burrows, where two or three young are born. These mice eat grass roots, tubers, and peanuts or roots of farm crops. They occasionally cause damage to trees by girdling the roots.

The red-backed mouse is found in the wooded mountains of the East and West and across Canada. It is about six inches long, with a short tail, and is brightly colored reddish brown on the back. This mouse is active day or night the year round, and it lives in the litter on the forest floor. It feeds on bark and other plant materials. These mice will kill trees by eating the bark around the trunk.

They build nests of grass on the ground in protected places, such as under rocks or logs.

Our recent experiences with mice range from annoying to amusing. They have also taught us it is almost impossible to mouse-proof a house. Our New Jersey home is a ten-year old well-constructed building, yet if we store bird food in a bag in the basement, in a very few days we attract mice. They quickly find their way into other parts of the house. Storing the seed in cans with a tight top eliminates the problem, and traps quickly eliminate the mice.

But in a house that is occupied for only short periods

of time, such as our farm, you learn to live with mice. You also learn what to look for when you arrive at the house after being away for a couple of weeks. White-footed mice are the problem. We never have seen them during the day but on occasion have seen them scurry across the floor·at night. Many times we have seen the results of their nighttime activity.

Once we arrived just after dark on a fall Friday and immediately started a fire in the old wood stove in the kitchen. Within minutes the kitchen was filled with smoke. After opening all the doors and windows and waiting half an hour for the smoke to clear, we found the cause of the problem. Mice had built a nest in the chimney.

On another occasion we had put on water to boil for a cup of tea. As the gas flared, there was an unusual smell. Investigation showed the mice had built a nest in the gas stove right next to the burner.

Now we have a checklist of things to look for before we light any fires or the gas.

One of our more interesting experiences was the year we had a very heavy infestation of mice. That year we had made the mistake of storing fireplace wood in the house over the winter, and that pile of wood was a huge mouse house.

We set three snap traps a night and before midnight caught three mice a night for several nights in a row. The easiest thing to do with the dead mice was to toss them out the back door on another woodpile to be picked up the next day. But the next day the mice were gone.

This went on for several nights and we were becoming thoroughly bewildered. What was taking the mice?

One night we rigged up a light with red filter and placed

the mice directly underneath. Very soon after we returned to the house we had the answer. In the dim light we saw a weasel make three trips from the woodpile to the mouse pile, taking the mice back to its den.

Combining the amusing with the annoying aspects of having mice around was the time my wife bought two new tufted bedspreads and placed them on the beds. In three weeks, while we were gone, the mice ate the tufts off the bedspreads and built a nest in one corner of the bureau drawer in each bedroom. So now we have untufted bedspreads.

We frequently find birdseed from the feeder, wild seeds from wild cherry trees and box elders carefully stored in the folded sheets in the linen closet or in the folded tablecloths in a cabinet. We've learned to look inside shoes before putting them on and to check boots, which are stored upside down but which have been known to contain mice nests or seeds.

The same year we had the large mouse population because of the woodpile, we had an old and lazy cocker spaniel. The dog's favorite resting spot was on a small rug halfway between the woodpile and the dining-room table.

The mice had learned that they could find tidbits under the table, so they ran from the woodpile to the table, in the process scampering over the outstretched legs of the sleeping dog. This awakened the dog, but by the time his eyes opened the mice were gone. If dogs can be bewildered, that one was. After a dozen such awakenings the dog moved elsewhere to sleep.

An artist friend who has a cottage not too far from our farm made a very interesting discovery about white-footed mice. He had live-trapped some and kept them in cages so he could study them for some of his artwork. The mice bred in captivity, and two out of one litter of young were very odd.

They slept all day in a small box nest and came out at dusk to feed. After feeding on dry cereal and seeds, they rested for a while, then proceeded to turn somersaults or flipflops for a few minutes. They would rest, and then somersault again. But the truly amazing thing was that one turned front somersaults and the other back somersaults while three or four inches up in the air above the floor of the cage.

Our friend checked with an animal behavior expert and found that this was not unusual but not common either.

Some species of mice, especially white-footed and deer mice are reasonably easy to trap alive for close observation and then release. The best traps are sold by scientific supply companies and cost three to four dollars. But one you can make from a mouse trap and tin can is shown in the illustration.

Bait either kind with peanut butter, cheese, or bacon grease, and set it in the basement if there are mice in the house or along walls or fences in an empty field, along woodland edges or close to the house in a suburban yard. Check the trap frequently, study the mouse for a minute or two, release it. If you want to keep it as a pet, be sure to have a good cage on hand before you set the trap.

An old aquarium or plastic terrarium with a tight top makes a good cage. Mice in captivity will eat birdseed, cold cereals, collected weed seeds, crumbs, or bits of crackers and bread.

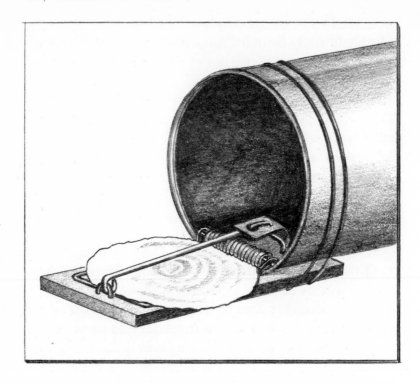

Mice can be amusing, annoying, or interesting depending on your point of view and personal experience with them. They serve a useful place in nature's scheme by providing food for foxes, hawks, owls and other animals that might eat more desirable forms of life. Predatory animals such as hawks, owls, foxes, weasels, or skunks will eat what is easiest to catch. When mice are easiest to find, they will usually not raid chicken houses or go after songbirds or rabbits.

3 Tricks of the Trade

There are several ways to discover which mice or other mammals live in your part of the country.

You can walk through fields and woods and never see a mouse, for they are active chiefly at night and stay hidden during daylight hours. But find an owl roost in a stand of cedars, pines, or spruce, and you can tell what kinds of mice live nearby and also that the mouse population may be very high.

How to do it? Look on the ground under the tree for what are known as owl pellets. These are furry-looking things about the size and shape of your thumb, larger or smaller depending on the species of owl. They are the indigestible parts of mice that an owl spits up.

Collect as many as you want and take them home for study. If they are wet, they are easy to pick apart with two large needles. If they are dry and hard, soak them in water to soften them. Pick them apart carefully, and separate the bones and skulls. Out of twenty pellets I collected once, I found the remains of twenty-two mice.

Many times you may never see an animal, but you can

find unmistakable clues that show an animal has been there, usually at night. Following are some of the mammal signs to look for.

Tracks

Skilled outdoorsmen can look at a track in snow, mud, sand, or soft earth and almost always tell what kind of mammal made it. It takes a great deal of experience and many hours in woods and fields to be able to identify most tracks you find. But with a little practice and a keen sense of observation, anyone can learn to identify the tracks of some of the more common smaller mammals.

The first thing to look for is the general pattern of the track, not just the print of one foot. The overall size of the entire track—the prints of all four feet—gives you a clue to the size of the animal and what kind of animal it is. The distance between front and hind feet and between hind feet or front feet is important to note. Does the track indicate that the animal walks or hops? You can usually tell by noting whether the tracks of the front and hind feet appear in pairs or alternately. Rabbit tracks, for example, show that a rabbit hops as it moves about. Fox tracks show that the animal walks.

Next, look closely at prints made by both front and hind feet. They will be different with the hind foot usually being larger. The size of these individual prints, their shapes and general form give you the first clue to the size of the animal and then to what kind it is.

Details of the print are important. Try to discover how many toes the animal has on its front and hind feet. See if there is an impression of toenails. These help you pin down the species of animal.

Most of the time you will have to identify tracks more by general pattern than by any other feature, unless you find fresh tracks in mud. Usually, though, the tracks you find in snow or sand are distorted and make the track look much larger than it really was when the animal made it.

Tracks can tell you more about mammals than just what they are.. They can show you whether the animal walks or hops, whether it was moving fast or slowly, and by following tracks, you can sometimes discover what the animal eats. I have followed fox tracks in the snow several times to see where the animal stopped to dig for a mouse. Raccoon tracks along a stream bank or lakeshore can lead to where the animal stopped to try to catch a crayfish. Deer tracks in snow can lead to the shrub or tree on which the deer browsed. But don't be like two hunters who once asked permission to hunt a bear on my farm.

We had arrived there the day before Thanksgiving to spend the weekend. About the first thing we noticed were bear tracks that came in the front gate, went around the house, out the back gate, and across a field. The tracks looked tremendous since the snow had melted considerably since the bear left them. This always makes tracks look larger than they actually are.

In any case, two hunters walking up the road saw the tracks at the gate and came to the house to ask if they might hunt. We said they could. Their problem was that they didn't know the front from the back of the track

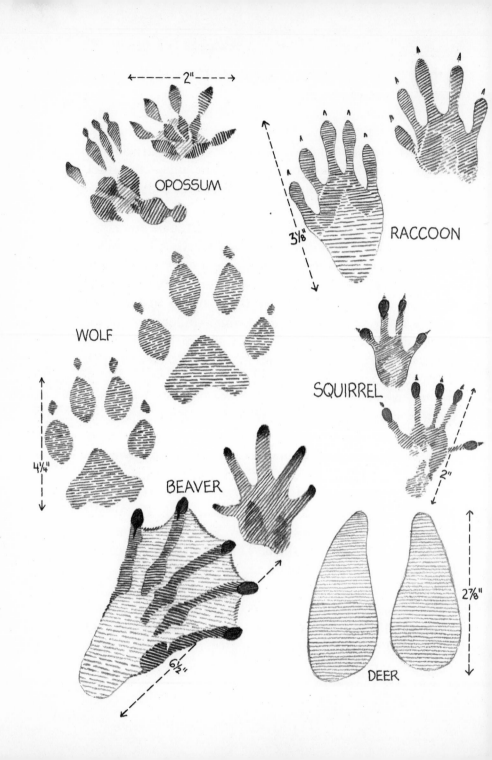

OPOSSUM

RACCOON

WOLF

SQUIRREL

BEAVER

DEER

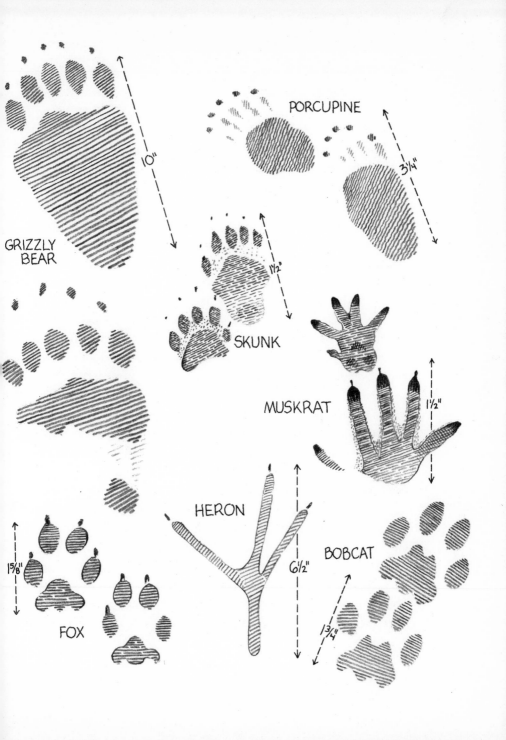

GRIZZLY BEAR — 10"

PORCUPINE — 3¼"

SKUNK — 1½"

MUSKRAT — 1½"

HERON — 6½"

BOBCAT — 1¾"

FOX — 1⅝"

and followed the tracks to where the bear came from and not where it was going.

Tracking Pit

Another way to find out which mammals may be active at night is to make one or more tracking pits. A tracking pit can be made almost anywhere from a suburban backyard to farm field or woods. Simply clear the ground in about a ten-foot circle down to mineral soil. Then dig up the soil to soften it and rake it smooth. Then place bait in the center, food such as cracked corn, pieces of apple or potato, pieces of bacon, sardines, lettuce or cabbage leaves, and some salt. The next day look for tracks in soft soil, and you will know what animals were feeding. If no animals came the first night, don't be discouraged. It may take a few nights for them to find the food.

Where this has been done in suburban New Jersey, white-tailed deer, raccoons, opossums, rabbits, white-footed mice, and chipmunks fed at one time or another over a week's time.

Lighted Tracking Pit

An even better way, but more complicated, is to rig up a red light over the pit with hundred-foot-long wire running to a hot-shot battery and dimmer switch. You can then sit hundred feet from the pit, focus binoculars or a telescope on it, and wait. After an hour or two or about 10 P.M., turn the switch to its lowest setting so that the light barely glows. Then very slowly adjust the switch so that the light gradually becomes brighter and brighter.

Many mammals are blind to red and pay no attention to the light. But you can see under a red light and thus can watch animals feeding on the bait.

A similar project is to make a small platform or shelf out of a board or piece of plywood and place it on a tree trunk or post about two to three feet up from the ground. Make the platform about eighteen inches wide and two feet long. Bait the platform with nuts, peanut butter, seeds, and small chucks of bacon. Place a red light over it with a dimmer switch and battery fifty feet away. In the early-evening hours it is often possible to attract deer mice, red-backed mice, field mice, flying squirrels, and other small mammals to such a feeder. When these mammals are attracted, sometimes weasels and other flesh eaters will be attracted, too, to feed on the rodents that feed on the bait. It may take several nights for them to find the bait, but they usually do sooner or later.

Track Prints

Sometimes you may find tracks that you cannot identify in a place that is too far away to go back to again. You may find some good sharp prints that you want to save to use again later on for identification. Or someone may doubt that a certain mammal lives in your area and you want to show track prints to prove it. It is easy to make a collection of animal tracks by making plaster casts out of plaster of paris or some similar material. You can keep these casts for many years, if you are careful, and in time build up a very interesting collection.

The equipment needed consists of a few strips of cardboard two inches wide and two to three feet long; some paper clips; a tin can; and a jar or can of plaster of paris. The plaster comes in a paper bag from the hardware or paint store. But the bag may tear or get wet, and you may lose the plaster. Keep it in a jar with a screw top or a can with a tight lid.

To make a cast of a single footprint, first make a hoop, slightly larger than the print from one of the cardboard strips. Fasten the hoop with paper clips. Place this hoop around the track in the mud or sand.

Then put about a cupful of plaster in the tin can, and add water, while you stir, until the consistency is about that of melted ice cream. Stir out all the lumps as you would in mixing pancake or cake batter, but work quickly. Plaster hardens very quickly.

Carefully pour the mixture into the cardboard ring so that the mixture runs into the track, filling every little crevice and indentation. Wait ten to twenty minutes before picking it up. Then wrap it carefully in old newspaper, and take it home.

Gently brush sand or mud and dirt off the cast, and get it as clean as you can.

The next step is to smear some oil or Vaseline on the print side of the cast. Make another hoop of cardboard so that it fits snugly around the cast. Press the hoop into soft soil with the print side of the cast up. There should be about an inch of cardboard extending up above the cast.

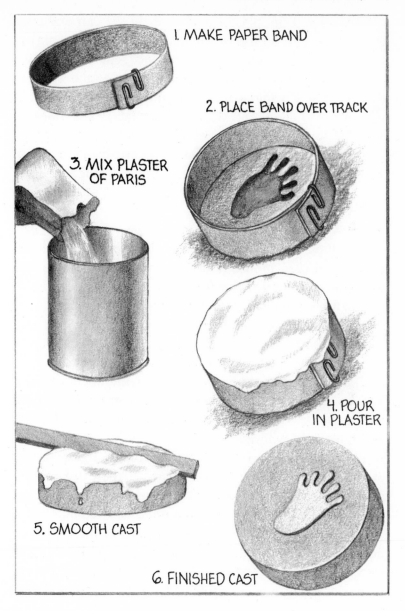

1. MAKE PAPER BAND

2. PLACE BAND OVER TRACK

3. MIX PLASTER OF PARIS

4. POUR IN PLASTER

5. SMOOTH CAST

6. FINISHED CAST

Mix some more plaster and water in the can and pour it into the mold on top of the first cast. When the plaster hardens, remove the cardboard, and separate the two casts. The second one will resemble the actual print in the ground. To make it a little more realistic, paint it with white shellac, and while the shellac is still "tacky," sprinkle some sand or dirt which has been sifted through a screen. When the shellac hardens, brush off the loose sand, and the cast will look exactly like a track in earth or mud.

Mammal Houses

You have probably heard of birdhouses, but have you ever heard of a mouse house? Some kinds of mice will live in houses, and it is a lot of fun to build a couple of mouse houses and set them out in appropriate places.

Deer mice or white-footed mice will occasionally move into a bluebird or wren house, so that gives us a clue to the size of the mouse house and where to set it out.

The house should be about four or five inches square and six inches high. The entrance hole should be about two inches in diameter. Any sort of box made of wood and unpainted will do.

Place it in a tree or even on a stump in the woods or along the edge of the woods. Leave it there for a week or two before inspecting it. If it is filled with grasses and leaves, the chances are a deer mouse has moved in.

Wooden boxes placed on the ground in high grass in fields or meadows frequently will be occupied by meadow mice. Make or get some small wooden boxes, and place them on the ground in protected places along field bor-

CHEESE BOX FOR
DEERMOUSE

SQUIRREL HOUSE

RACCOON HOUSE

WREN HOUSE
FOR FIELD MOUSE

ders or in clumps of high grass. Inspect them occasionally, and it will not take mice long to move in.

Raccoons and squirrels, too, will live in houses if they are large enough.

Old nail kegs or boxes about eighteen inches to twenty inches square and two feet high are about the right size. The entrance hole should be six to eight inches in diameter and placed near the top edge of the box.

Squirrel houses may be placed in oak woods, parks, yards, or anywhere you know squirrels live. They should be placed quite high in a tree.

Raccoon houses should be placed in wooded areas near swamps, marshes, ponds, or lakes. They may be placed in the fork of a tree eight feet or higher from the ground and in such a position that the animals can enter easily.

It is fun to make houses for mammals. Many times you can watch mammals easily at houses like these and get a better look at them than otherwise might be possible.

Animal Photography

Taking pictures of wild animals at night is fun and not too difficult. You will need a camera with synchronized flash or strobe, a tripod, and a cable release. In addition, you will need a four-inch strap hinge, rubber bands, a ten-penny finishing nail cut to one inch long and filed round on both ends.

The illustration shows how to fasten the hinge to the tripod with rubber bands so that it opens and closes easily. Place the cable release through a screw hole in the hinge so that the closing hinge will press the release and operate

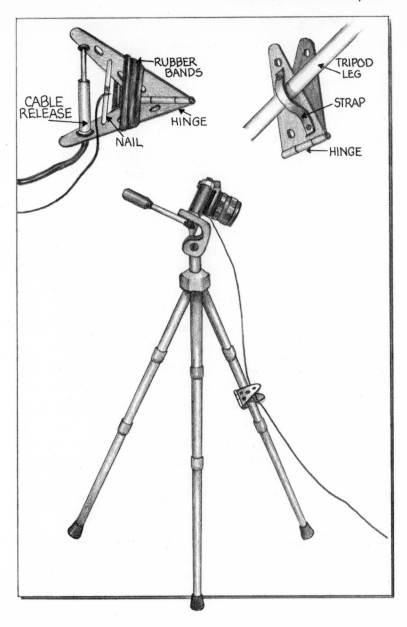

RUBBER BANDS

CABLE RELEASE

NAIL

HINGE

TRIPOD LEG

STRAP

HINGE

the shutter. Then loop three or four strong rubber bands around the hinge to hold it closed.

Open the hinge, and hold it open with the one-inch nail. Tie a piece of string or fine steel wire to the nail. When you pull the string or wire, the nail comes out of the hinge. The hinge closes and presses the cable release which trips the shutter.

Experiment with this gimmick for a while, and work out any "bugs" in it. I've used it for twenty years, and it always works; but each time I set it up, it takes a little adjusting here and there to make it work smoothly.

Focus the camera on the bait, set the camera to the appropriate f stop for distance and shutter setting, and back off or hide in a blind. When the animals come to feed, pull the string, and you have a picture.

Some camera stores sell remote-control devices that permit you to take pictures from thirty or forty feet away. They are not much more than long cable releases.

If you want to set a camera trap at a runway or water hole so the animal takes its own picture, you will need, in addition to the equipment described above, two wooden stakes about two feet long, some string, and a mouse trap.

Focus the camera on the runway, as before, set the shutter and lens, open the hinge, and place the nail in place. Only this time connect the string to the mouse trap that has been fastened to a stake driven in the ground as shown in the illustration. Drive the other stake in the ground on the opposite side of the runway as shown, and run another piece of string from that stake to the mouse trap. When the animal trips the trigger, the mouse trap snaps and pulls the nail out of the hinge, and the animal has taken its own picture.

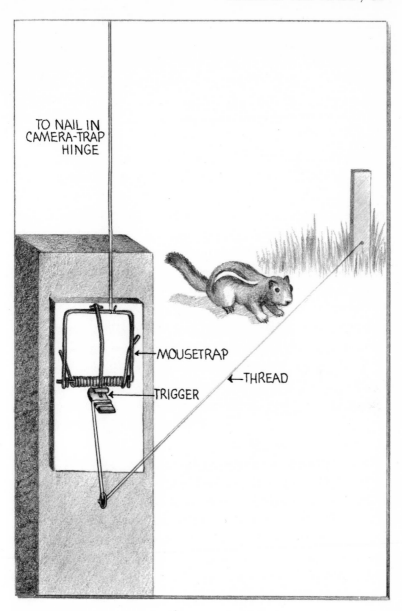

TO NAIL IN
CAMERA-TRAP
HINGE

MOUSETRAP

TRIGGER

THREAD

4 Birds at Night

Birds at night have made news headlines and have even attracted TV cameras each winter the last few years. When many million blackbirds congregate at night in a very few acres, the noise they make and the mess they spread can be not only annoying, but also a health hazard.

Generally, however, few birds are active at night except during migration in fall and spring. Then most birds probably migrate at night, especially the smaller birds.

The reason is that these birds—swallows, thrushes, sparrows, warblers, and blackbirds—need to see in order to find the insects and the seeds they eat. If they migrated by day, they would have to take time out to feed, and migration would be a slow process.

By migrating at night, they can feed and rest during the day and make much better time on the trip north or south. Also, by flying at night, small birds escape hawks that might prey on them. Hawks hunt by sight and consequently cannot see birds at night. Owls can see at night and hear very well and possibly do get a few birds. But owls feed close to the ground, and birds migrate at heights of five hundred feet and more.

The story of how birds can find their way at night, in fog or storm, with or without a moon or stars, is long, fascinating, and not yet completely understood. Entire books and long chapters in scientific publications have been written on migration.

One of the more dramatic stories that came out of one scientific study was that of a thrush. Scientists in Urbana, Illinois, had captured the bird and attached a miniature radio transmitter to a harness on its back. At a nearby airport there was a plane waiting with a receiver aboard that could pick up the signal from the thrush's transmitter.

The scientists hoped to follow the bird as it migrated north and determine how fast it flew and how far it would travel in one night. It fed and rested for a day or two, then at dusk one night took off on the journey north. The plane picked up the bird's signal and followed.

The plane lost the bird over Chicago because of radio interference but luckily picked it up again as it flew out over Lake Michigan. Sometime later the scientists saw a thunderstorm in the distance, and since they needed to refuel, they landed the plane with little hope of finding the bird again.

Again, luck was with them, and they found the bird had weathered the storm, which the airplane could not attempt. By morning the bird had reached the Upper Peninsula of Michigan, where it landed. In about twelve hours it had flown nonstop a distance of approximately the length of Lake Michigan and then some. That was something a small plane could not do.

You can get some idea of bird migration at night by focusing binoculars or a 20 to 25 power telescope on the full moon in August, September, and October. It is easier if the telescope is on a tripod and you are comfortably seated in a chair. It is best done when the moon is low in the sky since you are looking through more air at a lower elevation at which birds fly.

Remember that you are looking through a very small cone-shaped section of air compared to the vast space available for birds to fly through. Yet on a good flight night you may see a remarkable number of birds. It is virtually impossible to identify birds this way, except possibly for geese. The birds move quickly and may be a long distance away. On a good night I have counted eight to ten a minute for the three or four minutes I could look without taking a rest.

That most small birds fly at night and at low elevations is proved by the large number killed by flying into tall buildings, monuments, and even TV antennas. Special lighting has been developed for the Washington Monument and the Statue of Liberty to help birds avoid them. Some ornithologists make it a regular habit to search around such places after a foggy night. The birds they find are helpful in their research projects.

Another way in which scientists have tried to study birds at night is go to the top of a tall building such as the Empire State and listen. Some have even made tape recordings for later replay and study. Since some birds have a distinctive flight call, they can be identified by those who recognize the sound. In this way they can tell what kinds of birds are migrating at a particular time in the fall and get an idea of how many are in flight.

Other ornithologists have spent many nights in lighthouses along the shore where they can see birds briefly in the powerful beam of the searchlight.

During World War II in Europe, night-flying birds caused quite a disturbance in Britain several times. Radar was then in its infancy, and on one occasion a flock of geese set off all the alarm systems north of London. What seemed on the radar screens to be a fleet of bombers approaching turned out to be a flock of geese headed for their feeding grounds.

As higher-powered radar was developed, the problem increased. Even small birds and large birds at a greater distance showed up on the screens. It took considerable training to educate the radar operators so they could tell a bird pip on the screen from a plane.

Now much more efficient radar equipment is used to track birds and to learn at what altitudes they fly at night and how fast they fly. One goose was tracked for ninety minutes at a steady speed of thirty-five miles per hour.

Owls

Aside from migration, there are some birds active at night because they feed them. Owls are the best-known group. While owls can see very well in the daytime, the animals they feed on are most active at night. Snowy owls and short-eared owls are daytime feeders.

Like many nocturnal animals, owls may live close by without your knowing it. Barn owls nested within three-hundred feet of our back door, probably for several years, and we did not know it. There was an old abandoned house on the hill with all the windows but one in the attic boarded up, but that one gave owls a chance to fly in and set up housekeeping. It was not until the house was demolished that we knew about the owls.

Friends of ours, living in an apartment on the main street of a suburban Long Island town, occasionally heard a barn owl at night, but they could not locate a nest. Finally, they found it by discovering owl pellets on the ground. Looking up, they saw part of a nest under an overhanging roof.

Other friends never dreamed they had a screech owl in the yard. They had put up a birdhouse for flickers that used it the first spring and then left. When one of our friends climbed up in the tree to take down the box for the winter, he found it occupied by the small owl.

I once saw a whet owl in the middle of Central Park, in New York City, that perched daily in a spruce tree at the edge of a sidewalk traveled daily by hundreds of people. The bird was well camouflaged, and possibly no one had ever seen it.

If you want to find an owl, of course the first thing is to go out at night and listen. Owls, like all birds, have distinctive calls easy to identify once you have listened to a good recording. If you are a good mimic, you may learn to imitate an owl and actually call the owl to you.

Owls, like some other birds, establish a territory around their nest from which they will try to drive away other birds of the same species. When they hear you call, they think it another male and come to start a fight for the territory.

If you cannot imitate a horned or screech owl, you might be able to get or make a tape recording of the call to play back on a portable tape recorder. Some birdwatchers have used this system very effectively to call in owls at night.

Another method of attracting owls is to "squeak" them in. Since most nocturnal owls feed on mice which they hear, they can be attracted by a squeaking noise. You can buy commercially made squeakers at about two dollars, or make your own. Two small pieces of highly polished and waxed wood can be rubbed together to make a squeaking sound, or you can do it by loudly kissing the back of your hand.

My wife and I succeeded in attracting barred owls one night with a homemade squeaker. About 10 P.M. we heard a barred owl not far from the house. After going outside and standing under an apple tree, we squeaked and squeaked until there were two owls about six feet over our heads. We couldn't see them, but we certainly could hear their *whoo, who, whoo, who*. Owls fly silently, but you can hear their calls on a quiet night for a mile or more.

One summer night when I was in California, a friend showed me how to squeak in barn owls so that we could actually see them. He knew where they nested in an abandoned water tower, near which there was a bright streetlight. He placed a toy mouse he had bought in a novelty store on the ground under the light. He tied a thread to the mouse, and then he and I stood just outside the light. While he squeaked, he twitched the thread so that the mouse moved. On about the fourth squeak and twitch, down came a barn owl to grab the mouse. This worked about four times before the owl lived up to his reputation for wisdom and got smart to the trick. But my friend told me it would work about once a week.

Woodcock

A bird to listen for in late March or early April in the eastern half of the country is the American woodcock. This is a chunky, short-legged, long-billed bird a little larger than a robin.

Woodcock nest most commonly in a swampy thicket of alders, birches, or red maples or on the edge of a wet woods near an open pasture, field, or clearing. They need a good supply of earthworms, their major food, and thus the need for soft earth in the boggy thicket. The open field or pasture is necessary for their courtship flights, one of the more spectacular sights and sounds in the world of nature.

The evening song flight takes place during the time the female is laying and incubating eggs and starts soon after sunset. On dark nights it ceases when the afterglow disappears, but during the full moon it may last most of the night. It starts again at dawn and may continue to broad daylight. It takes place over an open area within hearing distance of the female.

The ceremony starts with the male walking around the nest strutting with tail erect and spread out and his long bill pointing down. While doing this, he produces a rasping *peent* sound which resembles the call of a nighthawk.

Suddenly he jumps into the air and flies upward, circling higher and higher in spirals until almost out of sight. During this upward flight he whistles continuously with twittering notes. Then he returns to earth, fluttering downward, circling and zigzagging, and finally gliding to the ground near the starting point. All during the downward flight he sings a musical three-syllable note that has been described as *chickaree, chickaree, chickaree.*

He then begins again with the *peent* notes and repeats the whole procedure. This is the only time of year, and a short time too, when the woodcock forsakes its secret ways and deliberately comes out in the open. The rest of the year it is seldom seen, unless you seek it out with a good dog or accidentally 'jump" one while walking through wet, cutover woods.

The first year we lived where we now do, we heard a woodcock nightly for a week in what was then an open field behind our house. A careful search in the wet woods turned up a nest with the female incubating eggs. Females on the nest are truly camouflaged and seem to know it. They remain motionless and will not flush until you almost touch them. In fact, some naturalists claim that they have actually petted a nesting bird without its taking off in flight.

The woodcock is unique among birds in the structure of its bill. The outer end of the upper mandible is flexible and can be moved away from the lower mandible when the bill is closed at the base and inserted in the ground. Thus, the woodcock can extract earthworms with a natural set of tweezers that surpass any made by humans. Woodcock locate worms either by hearing or through some sense of feeling in their feet. They frequently catch a worm on the first try. Looking for holes in the ground is another way of determining whether or not woodcock are in the area.

Goatsuckers

In addition to owls, there is another group of birds that is almost strictly nocturnal. The name "goatsuckers" was given to this group by the early European settlers because of the close resemblance to a group with that name in the old country. The group includes nighthawks, whippoorwills, poorwills, and chuck-will's-widows. The last three are named after their call, which is heard at night.

These birds have large tails, long, narrow, pointed wings, small bills, large mouths, and tiny feet. During the day they rest horizontally on tree limbs, fence rails, or the ground. At night or at dusk they capture insects on the wing.

Nighthawks commonly nest in cities on the flat roofs of buildings. They build no nests but lay their eggs on the gravel. Their call as they fly and feed is a *peent,* and that, and a broad white patch on each wing are the distinguishing characteristics. Even after dark when you cannot see them, the call is sure identification once you know it.

The name "nighthawk" is unfortunate. The bird is not a hawk but in general form resembles a small hawk.

Early one spring evening when I had an hour or so to wait for a train in Chicago, I walked around a block and heard nighthawks. Looking up, I saw a dozen or so swooping and diving after insects about a hundred feet above the ground. After watching them for a couple of minutes, I found myself almost surrounded by some be-whiskered, skid-row type of men, also looking up.

"Whatcha lookin' at, mister?" asked one.

"Birds," I said.

"Boids?" he replied. "We thought you saw the flying saucers we saw last night."

In the wild, nighthawks nest on bare soil or sometimes on flat rocks. They do not nest in colonies, as some birds do, but in late August they gather in huge flocks for their migration south. It is then that they make a spectacular sight at dusk as thousands dive and swoop for insects, feeding furiously to store up energy for their long trip.

One of the more spectacular sights in nature enjoyed by our family was the evening we stood on a rock outcrop, one hundred feet above the north shore of Lake Superior in Minnesota. For fully an hour we watched thousands of nighthawks fly by, feeding as they went, some below us close to the water, some at eye level, fifty feet away, some just over our heads. Below us they were dark-colored, but at eye level and above they were golden-colored as they reflected the light of the setting sun. You can see a similar sight if you are in the right place at the right time.

Since the major part of the insects eaten by nighthawks are harmful, the bird ranks high in the list of those beneficial to man. Unlike bats, which locate insects with their echo-location ability, nighthawks sweep up insects with their large mouths. They take all types of insects from large moths and beetles to the smallest of flies and mosquitoes.

Some nighthawk stomachs examined by scientists contained more than fifty kinds of individual insects. One stomach contained 2,175 flying ants, for example, and others held from 200 to 1,800 ants. In the Middle West nighthawks have been known to eat grasshoppers and locusts, which are harmful to agricultural crops. In the South nighthawks eat large numbers of boll weevil, an insect harmful to cotton.

Other birds in this group have similar habits and differ chiefly in the part of the country in which they are found. The chuck-will's-widow is a bird of the East and Southeast. The whippoorwill is found in the eastern half of the country, north into Canada, and the poorwill is a bird of the West. Like their relative the nighthawk, they are active from dusk into the night, feeding on large numbers of insects.

At one time people thought nighthawks and whippoorwills were the same bird, and even now in some places the two birds are not clearly distinguished one from the other.

What makes the whippoorwill famous is its call, which it may repeat many times without seeming to stop for a breath. It calls only at night, and the accent is on the first and last syllables—*whip-poor-weeel.*

A favorite calling perch for this bird in some places is the ridge of a cabin in the woods or near wooded areas. If you are inside the cabin trying to sleep or even carry on a normal conversation, forget it. We discovered that the hard way.

One night in June my family and I were in a small cabin on a lake in Quebec listening to an interesting political broadcast on our radio. The reception was not the best, nor was our understanding of French, but we were making out. Then along came a whippoorwill, which landed on the roof. After about one hundred *whips* in rapid succession, during which time we could not hear the radio, I banged on the underside of the roof with a broom handle.

The bird stopped for about two minutes, then started up again. Again I banged, and again it stopped. After going through that process four or five times, we gave up.

Naturalist Winsor Tyler, writing in a federal government publication, tells how he observed whippoorwills.

In order to study the whippoorwill at short range it is well to visit its haunts for a few evenings and learn how the bird we are to watch behaves when it wakes from its day's sleep. Whippoorwills move about over a considerable territory when they come into the open for their daily session of singing and feeding, they follow a route, evening after evening, that varies little, and on the circuit there are stations—a stone wall, a low branch, or a certain spot on the ground—where they are almost sure to stop and sing for a while.

If we seat ourselves near one of these stations where the light, which will be almost gone when the bird arrives, will favor our view, and where a dark background will obscure us from the bird, we shall be able to see the whippoorwill at short range, for if we sit motionless (no easy task, for mosquitoes will torture us) the bird will pay little attention to us. We must sit quiet and wait, following the song as it swings around the circuit, and we must watch the spot where the bird is about to alight, for, although in flight it looms big even in the dusk, when it comes to rest, with a flip of wings it becomes a bit of dead wood, a clod of earth, or vanishes altogether.

On several evenings late in May, at Wilton, N. H., I visited what appeared to be the whippoorwill headquarters—a dry wood of small deciduous growth bordering a sloping field, on one of which was a moist alder run that ran down to the edge of the wood. When I arrived, between sunset and dark, wood thrushes and veeries were singing, but before they quieted down for the night, the whippoorwills (from one bird to two or three) began to sing, always from the dry wood. They sang intermittently, and generally after each series of whip-poor-wills their voices came from a different part of the wood. By the time the light was becoming uncertain (when one would have difficulty in reading print) one bird, leaving the wood, worked up the slope, passing the field either by way of the alder run or by a wood of larger growth and an apple orchard that bordered the higher sides of the field.

On each of the first evenings when I visited the ground, one bird paused in the corner of the field where it joined the alder run, and sang a few times, and on two of these evenings I was able to approach the bird but not near enough to see it. The next evening, therefore, as soon as the bird that was singing in the wood began to change his position, I retired to this corner of the field to await him and sat down on a bank where my figure would not show against the sky. That evening was unusually dark and cloudy. The bird left the wood by the lower side, and at 7:50 I heard the song coming nearer and nearer through the alders behind me. Then, two minutes later, it came with startling suddenness from almost at my side. The bird sat on the bare ground at the foot of the bank not 6 yards from where I sat. In bringing my glass to bear upon him, I disturbed him, I think, for he flew silently away. He alighted, however, on a rock and began to sing. He was now 12 yards from me and on a level with my eyes. His side was toward me, and he faced nearly in the direction from which he had just flown. He sat flat on the stone with his head thrown slightly backward and upward and, on alighting, immediately began to sing.

Loons

One of the weirdest, if not the weirdest, sounds in the world of nature at night is the call of the common loon. Undoubtedly the term "crazy as a loon" came from its call which may be sounded while the bird is in flight or while swimming.

In Arthur Bent's *Life Histories of North American Diving Birds* the call is well described:

> The scream of the loon, uttered at evening, or on the approach of a storm, has to my ear, an unearthly and mournful tone resembling somewhat the distant howl of a wolf. It is a penetrating note, loud and weird, delivered with a prolonged rising inflection, dropping at the end, resembling the syllables A-ooo-OO, or as is often written O-O-ooh. Its laughter, however, is of a more pleasing quality, like the syllables hoo, hoo, hoo, hoo, hoo, uttered in a peculiarly vibrating tremolo.

Loons are birds of the north woods lakes from Minnesota to Maine and north into Canada. They are large birds with long, pointed bills. They dive for the fish they eat, swimming underwater as far as three hundred feet or more. They are clumsy on land but marvelous in the water.

Swifts

A bird active both day and night is the chimney swift, which some people mistake for a bat. It is a small bird that has been described as "a cigar with wings." As it swoops and darts through the air in pursuit of insects, it appears to flap its wings alternately instead of simultaneously.

Chimney swifts are found over a large part of the country. They are one of the few birds that have been helped by the settling and development of this continent. Before man built his many chimneys, the birds nested in hollow trees in the wilderness. Chinmeys provide exactly what swifts need for nest sites, and their numbers have increased greatly.

Thus, they are birds of cities, suburbs, and villages, as well as open country. As they gather in huge flocks in the fall and start their migration south, they sometimes cause considerable worry. One fall evening I received a telephone call from a concerned man who told me that a huge flock of bats just flew down his chimney. What should he do? I told him not to worry. They were chimney swifts roosting for the night and would leave the next day.

But I could well understand his concern. I once saw a huge flock of swifts fly down a chimney on Long Island, New York. They resembled a large cloud of black smoke going into the chimney instead of coming out. It was an awesome sight.

Swifts are unique among birds in that they rarely, if ever, set foot on land. They feed on the wing, roost in chimneys, barns, or trees, and really have no occasion to land on the ground. Some swallows feed in the same way

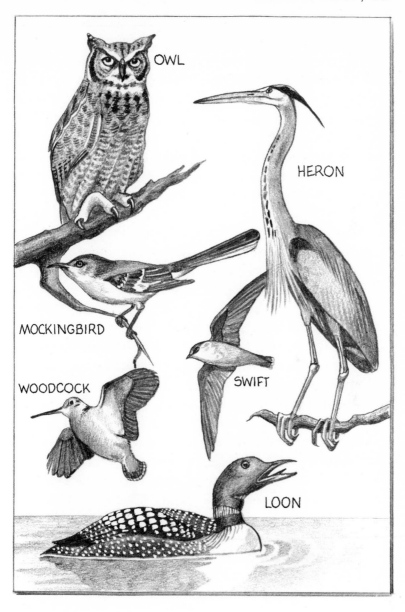

OWL

HERON

MOCKINGBIRD

WOODCOCK

SWIFT

LOON

and nest in or on barns and houses. But they must land to get mud to build their nests. Swifts excrete a saliva that is used to glue sticks together to form the nest and to fasten the nest to the inside of the chimney.

Once the nest is built and the young hatched, the adult birds are active around the clock, catching insects to feed themselves and the young. That swifts did feed at night was not known at first. An early ornithologist had a chimney in his bedroom and heard considerable activity at night. He found that the birds came and went for most of the night.

Naturalist Clarence Cottam describes very well the nighttime activity of swifts around the Capitol in Washington, D.C. The floodlights on the Capitol attracted insects, and the insects attracted birds.

On the nights when flocking occurred at the capitol, the birds began to arrive in small groups from all directions about sundown, and by the time they normally would have been going to roost they had formed into one great swarm. For the first fifteen or twenty minutes after sundown the birds foraged over the tree tops and flew in all directions without any apparent system to their movements, except that they remained in a rather restricted area. Gradually, as it grew darker, a greater number were seen to fly more or less in the same general circular direction; in other words, there was a distinct impression of group movement. About the time the lights came on or shortly thereafter, all were following a definite course. Each time flocks of incoming birds disrupted the rhythm and unison of the concentric flight there was a momentary disbanding. When they re-formed, however, all seemed instinctively to fly in the same direction. Most often the flight was uniformly circular, but occasionally it took the form of a conical cloud somewhat resembling a cyclone funnel. On one occasion it was seen to form a great figure "8" with one loop at a lower elevation than the other.

Herons

Yellow-crowned and black-crowned night herons are birds of freshwater and tidal marshes. They are long-billed birds about twenty inches high, short-legged for herons, with long, broad wings. They are more active at night, than in the daytime. Generally during the day they rest in trees or shrubs, flying out at dusk or even after dark to feed on small fish, crabs, or frogs.

We once lived in an apartment in a New Jersey suburb located on a small stream that ran the length of the town. The apartment complex stretched for about two blocks along one stream bank, and private houses lined the opposite bank. It did not seem a likely place to see black-crowned night herons.

One summer night, however, we came home late and heard the typical *quack* call of the heron. The call seemed to come from the stream, so we walked over to investigate. In the reflection of a nearby streetlight in the water we could see two herons stalking minnows and other small fish that lived in the stream. A few nights later the birds were in the same place, and we saw them several other times in the course of the summer. Although, we had seen night herons many times in marshes or along the shore from Florida to Maine, that was the first time we had seen them feeding right in town.

Some years ago a colony of thirty to forty roosted for much of the year in a half acre stand of pine trees in the middle of Cedarhurst, a town on Long Island, New York. Despite the fact that they are noisy birds, especially when they leave the roost, very few people knew they existed. They were hidden from view in dense branches and generally quiet during the day. They left the roost just after dark each night to fly a mile or so to marsh to feed. An apartment house now stands on the site.

Mockingbirds

A bird common to backyards and city parks over much of the country is the mockingbird. This bird is noted for its own song and ability to mimic other birds. John Burroughs said the bird was "famed for its powers of mimicry, which are truly wonderful, enabling the bird to exactly reproduce and even improve upon the notes of almost every other species of songster."

While mockingbirds sing at any time of day, they will sing well into the night when the moon is bright in spring and summer. They are also fond of starting the day early with full bursts of song shortly after dawn. These habits do not always endear the bird to the person who likes to sleep late or get to sleep early, especially since a favorite singing perch is the rooftop TV antenna.

Our home is near a high school athletic field, and from late August on we become accustomed to the frequent whistle blasts that come from the football coach as the team drills for the next week's game.

One September evening at just about dusk we were surprised to hear the whistle, not once but several times. We commented that the coach was going a little far in keeping the team working right through the dinner hour. Going out in the yard to investigate, we found that a mockingbird had learn to whistle so well that it sounded more like a whistle than a whistle did.

Another amusing story of mockingbirds comes from a friend in Florida: "It seems that Edward Bok, who created the well-known Singing Tower near Lake Wales, had several nightingales imported and confined there in cages. When the strangers had settled down and had begun to voice their famous song abroad across the orange groves, great satisfaction was felt, of course. Before long, however, nightingale songs were heard all over the surrounding territory! Here, there, and yonder the foreign strains were echoing, but all the captives remained in their cages. The mockingbirds of the area had taken charge and were broadcasting nightingale melodies over the countryside! It is said that the European performers were put to silence and soon refused to sing at all."

Almost anywhere you live there are birds to see or hear at night.

5 Waterways at Night

Fish

If you live on the southern California coast, you probably know about the famous grunion runs that take place in the spring. If you haven't heard about these fish, their story is one of the most interesting in all of nature at night.

On the first three nights after the full moon in April, May, or June, some of the beaches in southern California may be covered with six- or seven-inch-long grunion. What happens is this:

The fish swim into shallow water and wait for the tide to reach its peak. They then let the waves throw them on the shore in small groups. A female sinks her tail in the wet sand until less than half her body is exposed and lays her eggs. Males then fertilize the eggs. All this happens in between waves. One wave washes the fish on the beach; the next wave washes them back into the ocean.

The eggs remain in the sand four inches or so deep for two weeks, while the tide recedes and then comes back higher than before. The eggs will not hatch until the pounding of the waves stimulates hatching. When the wave action is right, the young fish crack through the eggshell, wriggle upward through the sand, and are washed out into the ocean.

Since grunion are good to eat, thousands and thousands of people visit the beaches during the nighttime spawning runs and almost go wild capturing the fish by hand in the short time they are onshore. If grunion spawned by daylight, they would not last long. Gulls and other birds would eat them by the millions and soon wipe them out. Thus, grunion are successful only on beaches where the highest tides occur at night.

Many other kinds of fish are active at night, as skilled fishermen will tell you. During the summer and fall nights, the coasts of southern New England, Long Island, New Jersey, California, and Oregon are lined with fishermen casting for striped bass. Some of the largest stripers landed by surf casting and weighing fifty pounds or more have been caught at night.

One of the best fishing experiences I ever had was on an early September night on the south shore of Long Island. The place was a foot bridge over a tidal creek, and the fish were small bluefish called snappers. A friend showed me how to do it.

The tide was coming in. This meant the snappers were coming in with it in search of small fish. My friend lowered a gasoline lantern on a line from the bridge until it was two to three feet above the water. The bright light attracted the bait fish, and the bait fish attracted the snappers. We dropped small, shiny lures ino the mass of bait fish and twitched the lure so it would flash in the light. Almost every twitch meant a snapper on the hook, and in less than an hour we had all we needed to feed both our families, with enough left over to freeze for another meal.

Some freshwater fish are active at night to the point where some bass fisherman go out only after sunset. In my youth I lived on salt water and had no experience with freshwater lakes until one summer when we camped on a southern Ontario lake. About eleven o'clock the first night, just as we were ready for bed, some loud splashes not too far offshore attracted our attention. There wasn't much of a moon, and we couldn't see what was splashing; but figured it must be fish feeding. Our rods were already rigged for early-morning fishing, so we attached a floating lure and cast it into the lake. The lure was the kind that made a gurgling noise when it was reeled in. On about the third gurgle a fish hit the lure, and for the next ten minutes I never had so much fun in my life. It was my first black bass, and it weighed a little over three pounds. My companions up and down the shore were equally successful, and our breakfast next morning was one of fresh bass.

The largest brown trout I ever caught was at night on a small stream in the Catskills of New York. It started when I was fly fishing one evening and saw a huge fish jump in a pool fifty feet or so upstream. I carefully waded to casting distance of the pool, but what I did not know was that the fish had moved downstream after the jump so that it saw me and quickly darted back to its shelter in the tree roots of an old yellow birch tree. Brown trout that size are smart fish. They have to be, or they would not live to that age.

First, I collected several dozen live grasshoppers. Then I wanted to find out where in the pool the fish fed after dark. Hiding in the willows at the upper end of the pool, I waited. As dusk blended into dark, I tossed some grasshoppers into the pool so they would float downstream. Nothing happened. I tried again. No results. About the fifth time I tried it, there was a loud splash at the bottom of the pool, telling me that at least a fish was feeding there, if not the monster.

The next day for an hour or so I stood downstream from the pool and practiced casting, so that in darkness I could drop a fly in the right place on the first try. I practiced for several days and fed the fish grasshoppers each night.

On the seventh night I thought I was ready. Entering the stream well below the pool, I inched my way upstream until I came to a flat rock that was my casting point. I used an artificial grasshopper fly.

The preparation and practice paid off. The first cast, which I couldn't see, was apparently perfect since the fly just about hit the water when the fish took the fly. Fifteen minutes later I had a six-pound brown trout in my net.

That story should give you a hint of a way to see fish at dusk whether you are a fisherman or not. Collect grasshoppers, crickets, caterpillars, or other insects, and throw them into a stream or lake. Some fish will jump clear out of the water; some will just suck the insect into their mouths. But in that split second you can get a good look at it.

Some of the biggest carp I have ever seen were in a New Jersey lake where a man who liked carp fed them each night at dusk. He took along a can of whole-kernel corn and from a dam tossed a spoonful into the lake. As the corn sank a foot or so, up came huge carp to feed, and they were in plain view for a second or so. Then the man rigged up his tackle, put a small hook in a kernel of corn, and tossed it and another spoonful into the water. He quickly hooked a fish and twenty minutes later landed a twenty-pound fish.

Another way to see fish is to walk along the shore with a bright flashlight. Keep the light off until you come to a likely spot. Then aim the light in the water. Many times bass and other fish will be feeding in shallow water, and you can see them briefly before they dart back into deep water. You can do the same thing from a boat as you row quietly fifteen feet out from shore.

Another interesting project is to drop a lighted waterproof flashlight into clear water that is about six feet deep. Sometimes the light will attract fish so they swim around the light. If you do not have a waterproof flashlight, use any kind, but protect it with a clear plastic bag that is folded over at the open end and tied securely with a rubber band or string.

To carry out another project that will show where fish spend the day and night in a lake, you must first catch a live fish or two. In addition to a cane pole, line, bobber, hook, and bait to catch a fish, you will need a fifteen-foot-long piece of nylon thread and a small balloon.

The best time to try this project is just before dusk when sunfish, yellow perch, and other fish are likely to be feeding along the shoreline. Before trying to catch a fish, blow up the balloon, and tie one end of the nylon thread to it. In the other end of the thread tie a simple slipknot noose.

If you are lucky and catch a fish, quickly slip the noose over the fish's head, and slide it back to the dorsal fin, the large fin in the middle of the back. Pull it tight, but not too tight. Then throw the fish back into the water. Sometimes the fish will swim out into the middle of the lake, and you can track it by watching the balloon. Sometimes it will stay along the shore to feed. In any case, until dark you can keep track of the fish, and the next day the balloon may still be attached and you can tell where the fish is.

After dark you may be able to pick up the balloon in the beam of a flashlight, as I have done many times. Some fishermen have used this technique to keep track of fish such as yellow perch that travel in schools.

Eels

One early fall night I visited a friend who lived on a small New Jersey lake that was formed by a dam across a small stream. There was one purpose for the visit: to see eels come out of the water and slither up to the dew-covered grass and make their way around the dam to reenter the stream below so they could ultimately get to the ocean. Eels can live out of water for some time, and the dozen or so we saw that night were out of water for perhaps an hour. They have been known to take twenty-four hours to travel from a land-locked lake to a stream or river.

The story of this nighttime migration starts in the Sargasso Sea, a very rich section of the Atlantic Ocean south of Bermuda. All eels from the Atlantic coast of North America and from Europe lay eggs in the Sargasso Sea. When the young hatch, they float around in the upper layers of the sea for about a year, feeding on very tiny plant and animal life. When they are strong enough to migrate, the European eels head for Europe, and the North American eels move into the Gulf Stream, which carries them northward.

They leave the Gulf Stream and swim west to reach our coastal rivers. Now two or three inches long, they swim by night and rest during the day. When they reach a river, usually the males stay in the lower part of the river, where there is some salt water, and the females continue upstream about as far as they can. From the river they swim up smaller streams until perhaps they come to a dam. Then, at night in late spring, when the dew is heavy they come on land and make their way to a lake or pond.

They stay in the lake until they are eight years old and quite large. When the time comes for them to make the long journey back to the Sargasso Sea, they are heavy with eggs.

If you are lucky enough to see eels on land in the fall, you are probably looking at females headed for a relatively small area of the Atlantic Ocean far away. If they are not caught in nets to be sold smoked as a delicacy, they arrive there in late fall.

Reptiles and Amphibians

A sure sign of spring over a large part of the country

is the dusk to after-dark chorus of the spring peeper, a tiny toad less than an inch and a half long. Despite its small size, the song of this toad can be heard up to a mile on a still spring night.

In the spring peepers live in small pools, marshy spots, and even roadside ditches. It is the male that calls so loudly. His purpose is to attract a mate. After the female lays the eggs underwater, the toads move to dry land, where they live on shrubs and trees and feed on insects. The eggs hatch in about ten days and in two to three months develop into tadpoles about an inch and a half long. They remain tadpoles until the next summer, when they develop into tiny toads. They then leave the water and live on land except when they return to mate.

Another very pleasant nighttime sound in spring is the trilling call of the American toad. This toad may be as much as four inches long but is usually smaller and is common in gardens and lawns where there are damp soil and places to hide.

In the spring males arrive first at shallow ponds and begin their nightly songs, mostly during rainy weather, to attract a mate. Once the eggs are laid the adults move back to a garden or lawn, where they do a great deal of good by eating harmful insects. They may be active during the day but are most active at dusk.

There are many other kinds of toads and frogs that liven up the spring night with distinctive calls. One of the first things we do when we arrive at our farm in April or May is to go at dusk and listen to different sounds we hear coming from an old pond not far from the house. Usually we can distinguish three or four different toads and frogs, and we have had fun making tape recordings.

The next day, as we look in the pond and around the edges we are amazed at the very few toads and frogs we see. But these few can be heard for half a mile at night.

One night I was driving along a narrow road in southern New York in a pouring rain when suddenly it seemed to be raining frogs and toads. Large numbers were hopping across the road from right to left. I stopped and watched, and in a minute or two only rain was falling again. I had read about these migrations but had never seen one before.

The animals were headed for a pond, and the urge to travel was great. They paid little attention to each other, sometimes colliding in midair as they hopped. But they had to reach the pond to mate and lay their eggs.

Some years later the director of a small nature center asked me to help him collect toads, frogs, fish, and other animals for his aquariums and other cages or tanks. We had two so-so days without much to show for our labors. The second night, driving home, I saw another of these migrations, this time of American toads. I jumped out of the car with a burlap bag in my hand and quickly collected a couple of dozen toads.

It was easier to return to the nature center than to make a special trip the next day, so I did. The director was away, and the safest place to leave the toads was in a small fenced-in pool near his house.

The next day he called. He was happy to have the toads, more than he needed, but he wasn't happy at being kept awake all night by the frogs singing under his bedroom window. A small frog can make a lot of noise.

6 Animals Without Backbones

So far only animals with backbones, called vertebrates, have been discussed. But many animals without backbones — insects, spiders, crustaceans, mollusks, and worms — are active only in darkness and are easy to observe.

Fireflies

Probably the best-known nighttime insect is the bettle called the firefly or lightning bug. There are at least fifty kinds of fireflies in the United States.

Scientists worked for many years to discover what it was that made the light of the firefly. They now know that it is two different chemicals reacting with oxygen to produce the cold light.

Recently chemists have developed a material that glows in the dark very similar to the firefly's light. It is called Cyalume and comes in a plastic tube. The pastelike material can be used to mark dark staircases or hallways. It is put on batons used by high school twirlers to make the batons more spectacular at night.

Fireflies are most active on humid nights and are found most frequently flying over damp meadows and fields. They are easy to catch by hand or with a small net. An interesting project is to catch one and put it in a closed jar. Place the jar in the open, and many times other fireflies will hover around the jar, flashing in time with the one inside.

Fireflies lay their eggs in damp places, such as under rotting logs or under decaying leaves. The eggs hatch into small wormlike creatures called larvae. The larvae also live in damp places. Since both eggs and larvae glow in the dark, it is fun to look for them. They are not easy to find, but it can be done.

Some young people in Florida found that by filling a jar with fireflies, they could use the light to read the large print in a newspaper, one word at a time.

Moths

Many kinds of moths are active at night, and you can see them flying under streetlights or near lighted windows.

There are two things you can do to get good close-up views of moths, and some of the larger moths are very beautiful. Make a simple frame of pieces of one-inch by two-inch wood about four feet square. Cover the frame with an old white sheet or similar material and stand it up on the edge of a woods or a fence row or even out in the backyard. On one side of the frame place a very bright light. It could be a 100-watt bulb on an extension running out from the house, or it could be a bright flashlight or electric lantern.

In any case, the light will attract moths, large and

small, and they will land on the sheet, where you can observe them.

Another way to attract night-flying insects is to smear molasses, honey, or a combination of the two on the leaves and branches of trees and shrubs. Do this during the day. At night quietly sneak up on the bait and light a flashlight. It may take more than one night for the moths to find the bait, but they usually do if you give them time.

Some flowers such as honeysuckle and evening primrose are attractive to moths such as the sphinx and clearwing moths. You may be able to see some of these interesting insects by checking flowers at night.

Other Insects

Other insects, such as several of the beetles, are active only at night, and there is one good way to see them.

Dig a hole in the ground in a wooded area or on the edge of a woods or near a brushy fence row. Place a two-pound coffee can or other deep can in the ground so the open end of the can is just at ground level. Fill in the hole around the can.

Place a small piece of raw meat or raw fish in the can. Wait a few days until the meat or fish starts to decay and look in the can. Beetles and other insects will go in to eat the bait but cannot climb out. You can study them carefully and let them go or keep them for an insect collection.

Mollusks

The shells you find along the beach or on rocks, mud

flats or sandbars are the "homes" of a group of animals called mollusks. The shells you find are usually dead. But you can see the live animals, after dark, and when alive they are much more colorful.

Wear sneakers and a bathing suit, and take a bright flashlight. Mollusks avoid light and during the day hide under the rocks or dig down in the sand or mud. At night they come out to feed. By walking carefully in a foot or two of water, you can see them and scoop them up with a small net or even a kitchen strainer.

The best place to look is in calm water on the edge of sandbars or mud flats or along rocky shorelines.

There are mollusks without shells that are found in many gardens, if it is a wet summer. Called slugs, they are fond of some vegetables such as tomatoes and lettuce, and you may find them by looking at night. During the day they hide under boards or in damp leaves or grass.

As it moves around, the snail gives off a slimy substance that will glisten in the beam of a flashlight. You may see the slug trail before you see the slug.

There are land snails with shells, too, that are active at night. Some live in trees, some on the ground, some in rotting wood. There are snails and other crustaceans that live in fresh water and are active at night. Their habits are similar to their salt-water cousins.

Crustaceans

Crabs of various kinds and crayfish in this group also become active once the sun goes down.

Hermit crabs are easy to find along some beaches and fun to catch and keep as pets. These are soft-bodied crabs that use a mollusk shell for protection. If you see a snail shell or similar shell moving across the sand or over rocks, the chances are good that it is a hermit crab.

Our family found out that hermit crabs are active at night after we collected six or seven kinds of shells and kept them in an aquarium on a windowsill. We fed them lettuce, celery leaves, and small pieces of toast. The first night we had them the light was out in the room where they were and we were in another room reading. About ten o'clock we heard some clicking noises coming from the dining room. At first we didn't know what it was. We turned on the light to look in the aquarium, and all was still. Ten minutes after the light went off, we heard the noise again. On went the light, and all was quiet.

We finally found that the hermit crabs were trying to get out of the aquarium, and were climbing on top of each other in one corner. The click sound came when the one on top fell off and landed on the one below.

They soon became adapted to captivity and made very interesting pets. In fact, we used our knowledge of the fact that they avoid light to make a good movie that shows how a hermit crab walks.

On one side of the aquarium we placed a flat board so it leaned against the side at an angle. This provided space beneath where the crabs could hide. They quickly found the hiding place and used it most of the time.

To make the movie, we placed a bright floodlight over the aquarium and took two snails out of their hiding spot and placed them on the other side of the aquarium. As they walked back to hide, we shot the movie close-ups.

On a vacation in the Virgin Islands recently, we used the same technique to film much larger hermit crabs, except there we made the movie under natural conditions. We caught the crabs at night on the rocky beach and kept them until the next day. Then we placed them in the open and filmed them as they walked to the shelter of a large rock.

Fiddler crabs live in holes in the sand just above the high tide mark along many beaches. They stay in these holes during the day and come out at night to walk around and feed. If you find a hole or holes about the size of your finger, remember where it is, and go back at night. With red plastic over your flashlight you can easily follow a crab and watch its nightime activities.

Crayfish, the three- to five-inch relatives of lobsters, are very sensitive to light and stay hidden under rocks or in the mud until the daylight dims. Then their activity begins as they feed on a variety of plant and small animal life. They must be wary, though, for bass and other fish may feed on them.

In the South one crayfish lives on land where water is close to the surface. It makes a hole down to the water using the soil it excavates to build a chimney over the hole. The chimney keeps the sun from entering the hole and drying up the water. The crayfish come out at night to feed on nearby plants.

Occasionally dozens of crayfish will migrate from one wet place to another to the bewilderment of motorists who see them in the headlight beams.

Spiders

Many kinds of spiders are nocturnal creatures and can

be found in many different places, from damp basements to empty lots in cities, parks, country roadsides, and open fields. The eyes of some spiders glow in the beam of a flashlight or car headlights and are easy to see. Move slowly and carefully, and you may find a spider on or near its web. Quickly cover the flashlight with red plastic, and you may be able to see the spider work on its web or capture an insect that flies into the web.

Worms

If you like to go fishing and use worms for bait, there is a way of getting them that is much easier than digging. Wait until after the dew forms at night, and use a flashlight with red lens or a regular flashlight with red plastic over the white lens.

Walk very carefully and quietly over a closely mowed lawn, and slowly move the flashlight beam back and forth. You can see the worms glisten in the red light, but you must be quick to catch them. If they are not completely out of the hole, they can get back in with remarkable speed. On a rainy night worms are especially easy to catch in this way.

If you do not need worms for fish bait, you can still catch some and put them in a worm house where you can watch them at night with a red light and see how they move through the soil.

Index

THE AUTHOR

Ted S. Pettit, Conservation Director of the Boy Scouts of America, is a prize-winning nature writer who has published thirteen books and innumerable articles. His two previous popular books from Putnam's are *Bird Feeders and Shelters You Can Make* and *The Long, Long Pollution Crisis*. Mr. Pettit, who travels widely in his nature studies, lives in Bridgewater, N.J., and has a vacation home in Westkill, N.Y.